Fluid Power
Educational
Series

Hydraulic Accumulators
and Circuits

(In the SI units)

Joji Parambath

Hydraulic Accumulators and Circuits
(In the SI Units)

Copyright © 2021 Joji Parambath

All rights reserved

ISBN: 9798653863189

https://jojibooks.com

First Edition – 2020
Revised Edition - 2021

Disclaimer of Liability

The contents of this book have been checked for accuracy. Since deviations cannot be precluded entirely, we cannot guarantee full agreement. Only qualified personnel should be allowed to install and work hydraulic equipment. Qualified persons are defined as persons who are authorised to commission, ground, and tag circuits, equipment, and systems following established safety practices and standards.

Table of Contents

PREFACE

Hydraulic accumulators are special devices that are extensively used in hydraulic systems to realize many interesting control functions. However, the functions, constructional features, and control circuits of accumulators appear to be a difficult proposition for hydraulic professionals who are not properly trained or a newcomer in the field of hydraulics.

The book brings out the essential technical information related to hydraulic accumulators extracted, especially from the materials available from the manufacturer's domain. The book explains the functions, classification, constructional details, and comparison of many types of accumulators including piston, diaphragm, and bladder types. Further, the book presents the topics on pre-charging, safety requirements, and applications of accumulators, in sufficient detail. A chapter gives some basic circuits of accumulators. Another chapter presents the topic of the sizing of accumulators with many numerical examples. Note that, the book uses the SI system of units. The topics of maintenance and specifications of accumulators are given at the end of the book.

The book is written with the main objective of providing all essential information about accumulators especially to untrained professionals and newcomers in the field of hydraulics. The topics are logically arranged for a simple to the complex-level progression of the subject matter for the quick understanding of the constructional features and circuits of accumulators.

Many other fluid power topics are given in other textbooks under the fluid power educational series by the same author. A list of all the textbooks is given at the end of this book (Page No. 61). Also, please see the details at https://jojibooks.com
Enjoy reading the book.
Your feedback is most welcome.

JOJI Parambath

Chapter 1 | Functions of Hydraulic Accumulators

An accumulator is a device used for absorbing shock pressures and storing energy in a hydraulic system. It mainly consists of a vessel in which a hydraulic fluid is held under pressure by a raised weight or spring or volume of compressed gas. It is, thus, possible to store potential energy in the accumulator, when the associated system pressure remains higher than that of the accumulator. The accumulator can release the stored energy back into the system for performing some useful hydraulic task when the system pressure falls below that of the accumulator. The employment of accumulators in machinery can improve the performance of the machinery, realize greater energy efficiency and reduce noise generation. This book explains various aspects of hydraulic accumulators, such as their functions, construction, classifications, safety aspects, maintenance, and sizing, and one chapter, later on, presents typical hydraulic circuits employing accumulators. Figure 1.1 shows the realistic view of a gas-charged accumulator with a safety control block.

Figure 1.1 | A hydraulic accumulator with a safety shut-off valve

Accumulators are employed in hydraulic systems to realise many special functions. The essential functions are presented in the following sections.

Generation of Shocks in Hydraulic Systems

Shock pressures are developed in a hydraulic system when the system flow is abruptly blocked, or the direction of flow is changed suddenly, as the associated directional valve is shifted. Shocks may also be developed from the jerkiness of the load attached to a cylinder or motor in the system. Next, it may also occur due to the application of external mechanical forces.

Shock Absorbing Function of Hydraulic Accumulators

Two circuits given in Figure 1.2, demonstrate the effect of shock pressures on a hydraulic system with a pump, valve, actuator, and interconnecting lines, and the elimination of shock pressures with the addition of an accumulator to the system. Figure 1.2(a), shows the movement of a column of high-energy fluid through the hydraulic circuit without an accumulator. When the flow is blocked abruptly, for example, with the rapid closure of the valve, pressure waves are developed in the system which travels back and forth through the system. The resulting hydraulic shock or 'water hammer', as it is sometimes called, and the vibration can develop peak pressures several times greater than the normal working pressure and may cause severe damage to the associated components of the hydraulic system. For example, the shock pressures may rip tubing, blow seals, produce fatigue failures, jar parts, and split the pump housing.

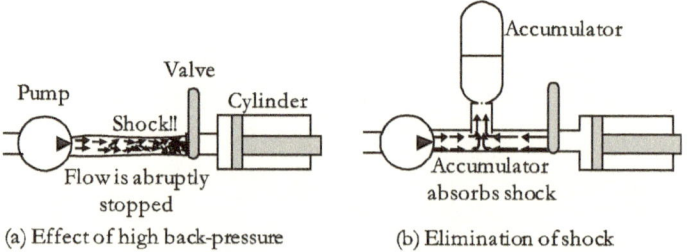

(a) Effect of high back-pressure   (b) Elimination of shock

Figure 1.2 | Schematic diagrams illustrating the development of shock pressure in a hydraulic system and its elimination

The damaging effects of the hydraulic shock can be prevented by installing an appropriately-sized accumulator in the system, as illustrated in Figure 1.2(b). The accumulator can absorb the kinetic energy of the moving column of the fluid and can suppress the resulting hydraulic shock. By this action, the system parts such as pumps, valves, hoses, and fittings get away from the pressure spikes. It may be noted that an accumulator is, usually, connected as close to the demand point as possible to overcome the flow restrictions and the drag from long pipe runs.

Other Functions of Hydraulic Accumulators

Apart from the functions of storing energy and absorbing shocks, as described in the previous section, accumulators can also provide many other functions, such as dampening pressure pulsations, storing energy during periods of low demand and releasing the energy during periods of high demand, and compensating leakages. For example, an accumulator can be used in an application requiring a large volume of hydraulic fluid for a very brief period. With the realisation of some of these functions, accumulators can reduce the pump capacity requirements, system leakage, and noise levels in hydraulic systems. Accumulators can assist a hydraulic system in reducing energy and maintenance costs and can improve system performance, efficiency, and reliability.

Pulsation Dampening

Pressure pulsations in a hydraulic system are, usually, caused by delivery fluctuations from the system pump, irregularities in the fluid flow, thermal variations, or excessive loads. The pressure fluctuations in the system can cause variations in the actuator's speed and inferior system performance. Adding a properly-located and correctly-sized accumulator to the system cushions the pressure pulsations and keeps the system pressure relatively constant. The cushioning of the pressure pulsations further leads to reduced noise levels in the system.

Energy Storage and Release

A typical machine cycle during the operation of a hydraulic system consists of extended periods (say, 80% of cycle time) of little flow and short periods (say, 20% of cycle time) of high-volume flow, as shown graphically in Figure 1.3.

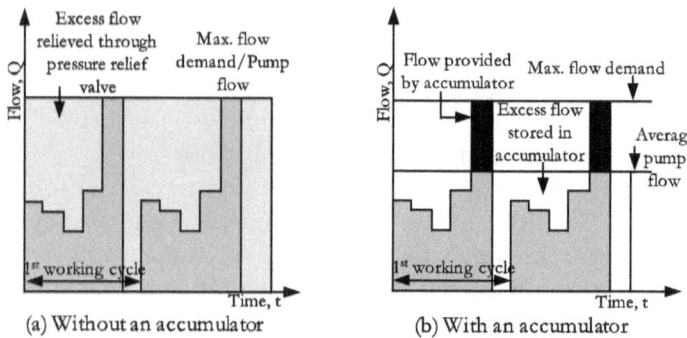

(a) Without an accumulator   (b) With an accumulator

Figure 1.3: | Typical machine cycles in hydraulic systems

If a fixed-displacement pump is used in the system with an intermittent duty cycle to supply the system fluid, then, the system pressure relief valve bypasses the fluid for most (say 80%) of the time. If the pump runs continuously, then the discharge of the pressurised fluid through the associated pressure relief valve represents a considerable loss of power, as shown graphically in Figure 1.3(a).

With the addition of a correctly-sized accumulator, energy can be stored in it during periods of low demand and the stored energy can then be added to the pump flow during short periods of high demand. Therefore, the pump unit in the system can be resized for the discharge of the high-pressure fluid to match the average power requirement of the machine's operating cycle, as shown graphically in Figure 1.3(b).

The result is the economising of the pump drive power. The use of the accumulator also results in a quick response to temporary peak energy demands. A rule of thumb suggests that an accumulator can be designed into the system if the pump is working at full load for less than 20% of the time.

Cost Reduction
The employment of an accumulator in a high-performance hydraulic system with an intermittent duty cycle permits the use of a smaller capacity pump in the system. As we are aware, an accumulator can store energy during periods of low demand. This energy can be released upon demand and is available for immediate use. In turn, the smaller capacity pump uses a lesser amount of fluid during the system operation, and this fact entails the use of a smaller reservoir. The net result is the cost reduction in the construction of the system.

Leak Compensation
On hydraulic presses used for moulding, bonding, etc., it is necessary to maintain constant pressure on the work during long curing periods. However, pressure changes can occur in such systems due to the leakage of system fluid. A charged hydraulic accumulator can compensate for the pressure changes by supplying additional fluid to the system while the pump remaining unloaded, during an extended idle period of the system, to make up for the loss of the system fluid.

Auxiliary Power Source
Energy stored in a fully charged accumulator can be availed in a hydraulic system to meet a sudden demand for a high level of power for a comparatively short time to complete the cycle or as a standby source of power in an emergency during power failures. It can also be used for the independent operation of auxiliary or pilot circuits in a hydraulic system when the pump flow is required to perform the main operating movements in the system.

Accumulators are not meant for every hydraulic system. They involve cost; therefore, they can be used if they offer an advantage over conventional circuits.

Chapter 2 | An Overview of Accumulators

This chapter presents the classification of accumulators and constructional features of weight-loaded and spring-loaded hydraulic accumulators. The constructional details of hydro-pneumatic accumulators are presented in subsequent chapters.

According to the method of construction, accumulators are classified into three basic types. They are (1) weight-loaded accumulators, (2) Spring-loaded accumulators, and (3) hydro-pneumatic (gas-charged) accumulators. The weight-loaded types are no longer used in modern hydraulic applications, and the spring-loaded types are now virtually confined to small mobile machinery. In modern hydraulic systems, the preferred type is the hydro-pneumatic accumulators. They are further classified as bladder, diaphragm, piston, and bellows types. They are available in a full gamut of sizes, pressure ratings, materials, and port configurations. The classification chart of Figure 2.1 shows all these types and their sub-types.

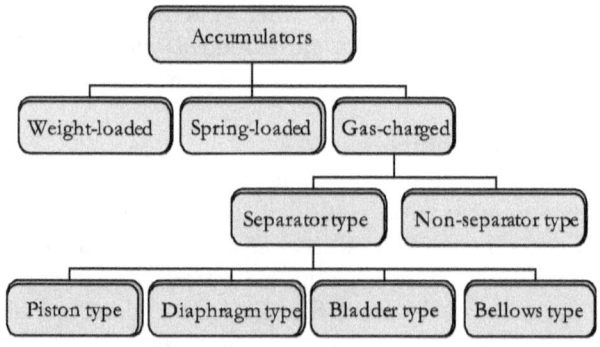

Figure 2.1 | Classification of hydraulic accumulators

General Constructional Features of Accumulators

Figure 2.2 shows the simplified cutaway views of the different types of hydraulic accumulators. They are not complete representations of accumulators, but they illustrate the general constructional features of accumulators for initial learning.

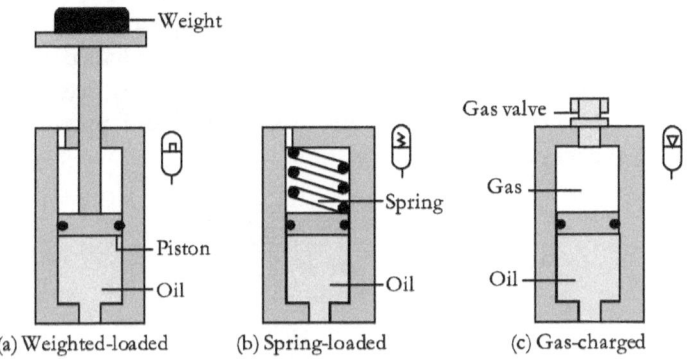

Figure 2.2 | Cross-sectional views of different types of hydraulic accumulators

An accumulator, usually, consists of two chambers that are separated by a piston, diaphragm, or bladder (bag). One chamber is meant for admitting the fluid from the associated system, and the second chamber is for maintaining a weight or spring or volume of pressurised gas. Hydraulic energy is stored when the system fluid under pressure works against the weight-loaded or spring-charged piston, or gas-charged piston, diaphragm or bag.

Accumulator Symbols
The standard symbols, shown in Figure 2.3, are used to describe the different types of accumulators as per the standard ISO 1219.

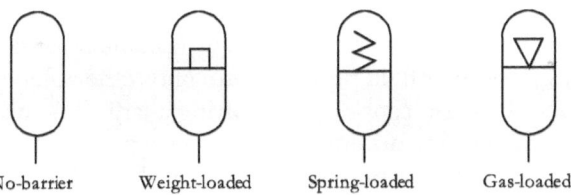

Figure 2.3 | Symbolic representations of hydraulic accumulators

Weight-loaded Accumulator

The weight-loaded hydraulic accumulator consists of a vertically-mounted, thick-walled steel cylinder with a piston. A deadweight or a series of dead weights is placed on the top of the piston, as shown in Figure 2.4. A correctly-sized port is provided for the hydraulic connection. The non-pressure side, usually, has a drain port (not shown) for relieving any leakage fluid and avoiding any potential backpressure on the piston.

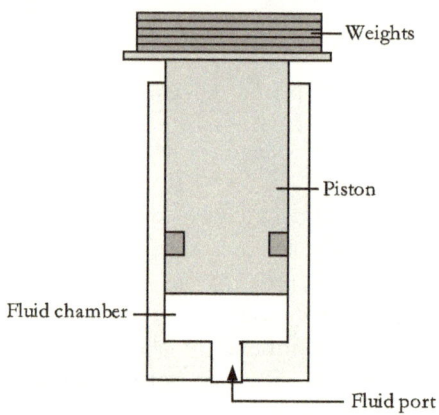

Figure 2.4 | A weight-loaded accumulator

The piston, along with the dead weight, is raised when the fluid under pressure works against the piston. At the same time, the weight exerts a downward push on the piston as a result of the force of gravity. As the system pressure drops, the fluid from the accumulator is forced out into the hydraulic circuit by the weight.

An advantage of the weight-loaded accumulator is that it produces a constant fluid pressure throughout the volume output of the unit. It is also capable of supplying a large volume of fluid under high pressure. Accumulators of this type are used in large forging and moulding presses. The weight-loaded accumulators are bulky, cumbersome, and expensive. Moreover, their portability is out of the question.

Spring-loaded Accumulator

The spring-loaded accumulator, as shown in Figure 2.5, consists of a cylinder body, movable piston, and spring. The piston in the chamber is pre-loaded with a spring. The non-pressure side, usually, has a drain port (not shown).

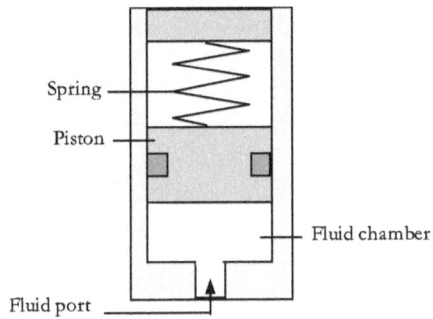

Figure 2.5 | A spring-loaded accumulator

As the pressurised fluid enters the chamber from the associated hydraulic circuit, the spring gets compressed. The compressed spring always acts against the piston. As the system pressure reduces, the fluid from the accumulator is forced out into the circuit by the charged spring.

The spring-loaded accumulators are, usually, smaller and less expensive than the weight-loaded accumulators. Their mounting is easy, and they can be fitted directly to the power unit. The pressure generated by this type of accumulator depends on the size and pre-loading of its spring. The accumulator pressure reaches its peak value as the spring is compressed to a maximum, and the pressure drops to its lowest level as the spring approaches its free length. Therefore, the pressure does not remain constant throughout the volume output of the spring-loaded accumulator. The spring-loaded accumulators are typically suitable for low-pressure, low-volume hydraulic applications, and they cannot be used in applications requiring high cycle rates.

Gas-charged Accumulators

The low compressibility of the hydraulic fluid makes it difficult to store hydraulic energy in small volumes, but, it can transfer a considerable amount of force to a load surface. On the other hand, gas is a highly compressible medium and can store large amounts of energy in small volumes. The gas-charged accumulators make use of these two properties for their superior performance as compared to other types of accumulators. Hence, they are, by far, the most commonly used type of accumulators. Moreover, they offer a good dynamic response. Remember, the design, manufacturing, and testing of accumulators should conform to relevant standards. Next, the typical design life for a hydraulic accumulator is 12 years.

An inert gas such as dry nitrogen should always be used as the gas medium. It may be noted that neither air nor oxygen should be used for charging an accumulator. If used, there is a possibility of auto-ignition of the gas medium at high pressures and hence there is a danger of explosion.

The gas-charged accumulators fall into two basic categories. They are: (1) Non-separator type and (2) Separator type.

Non-separator type Accumulator

Some of the earlier accumulators used in hydraulic systems were the non-separator type (or direct-contact type) fluid-gas containers. Approximately half of this type of accumulator was filled with the fluid, and the other half was filled with nitrogen gas, with no physical separation between them. The shut-off valve at the fluid port was required to be closed before stopping the system from preventing the escape of the fluid and gas from the accumulator. In this type of accumulator, there used to be foaming and absorption of the gas into the fluid medium. The gas used to get dissolved in the fluid and then carried away. Therefore, non-separator type accumulators are not used in modern hydraulic systems, but many such older units may still be in service.

Separator type Accumulator

A commonly used type of gas-charged accumulators is the separator type. In this type, there is a physical barrier, such as a diaphragm or a bladder (bag) or a floating piston, between the gas medium and the fluid medium contained in it. This barrier is designed to utilise the compressibility of the gas for absorbing the energy efficiently. Further, a correctly-sized port is provided for the hydraulic connection.

Pre-charging, Hydro-pneumatic Accumulators

A hydraulic accumulator must be pre-charged with clean nitrogen gas of at least Class 4.0 with a filtration level of 3μm or better. A fill port is provided to supply nitrogen gas. More details of pre-charging a hydro-pneumatic accumulator is given in chapter 8.

Other Features, Hydro-pneumatic Accumulators

The installation, maintenance, and safety aspects of hydro-pneumatic accumulators are also very important for any hydraulic system with accumulators and are discussed in the following sections and chapters.

Classification, Hydro-pneumatic Accumulators

The separator type of accumulators can be classified into the following three types, according to the type of barrier used to separate the hydraulic fluid chamber and energy storage section:

- Piston Type
- Diaphragm type
- Bladder type
- Metal bellows type

These types of accumulators are popular for high-pressure hydraulic applications. Each type has its advantages and disadvantages. The constructional details of these types of accumulators are given in the following chapters. A comparison of these types of accumulators is given in Table 7.1, Chapter 7.

Chapter 3 | Piston Accumulators

A piston-type accumulator consists of a cylinder (or shell) with a finely finished internal surface, a metal piston, end caps, a fluid port and a gas charging valve. The piston separates the cylinder into two chambers. One chamber is to contain hydraulic fluid on the system side and the other one is to hold nitrogen gas. The gas section is pre-charged with dry nitrogen gas. A schematic diagram of a piston accumulator is shown in Figure 3.1.

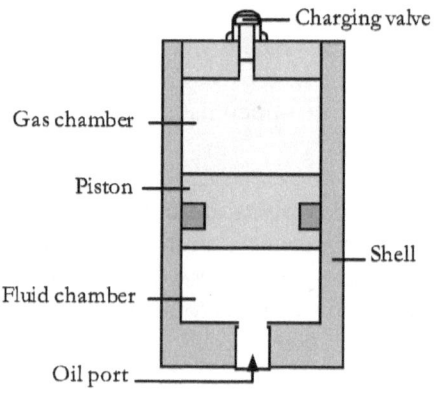

Figure 3.1 | A piston accumulator

Working, Piston Accumulator: The fluid chamber is connected to the hydraulic system so that the accumulator can draw fluid when the system pressure goes up, thus compressing the gas. When the system pressure drops, the pressurised gas expands and forces the fluid back into the system. That is, the compressed gas provides a quick pneumatic spring action to force stored energy from the accumulator into the system.

Shell, Piston Accumulator: The shell is manufactured from a piece of homogeneous, seamless tubing with a very finely machined internal surface. It is made of carbon steel, stainless steel, low-temperature carbon steel, or other materials. For use with a chemically aggressive fluid, the interior and the exterior of the carbon steel shell can optionally be provided with corrosion protection, with plastic coating using epoxy or chemical plating using nickel or chrome. Alternatively, stainless steel can be used for accumulator parts that are liable to be exposed to a corrosive environment or subjected to high pressure.

Piston: The freely floating and movable piston with a pair of O-ring seals and guide rings separates the inner hollow space of the cylinder into two chambers. The piston must act as a leak-proof separation element. The piston is typically made from aluminium, carbon steel, or stainless steel. The piston must be lightweight for its good acceleration. Piston diameters are typically available from 40 to 610 mm in a graded manner. Piston velocity is limited for various models from 0.5 to 5 m/s (typical).

Sealing, Piston Accumulators: The piston floats on the low-friction O-ring seals to prevent metal-to-metal contact between the piston and the internal surface of the accumulator. Seal materials meant for piston accumulators are formulated from the most advanced elastomers capable of meeting a wide range of temperatures from -40°C to +150°C. The advantages of using low-friction seals include no stick-slip, low wear, and high piston velocity (up to 5 m/s).

Seal materials must also be compatible with a wide variety of fluids including mineral-based fluids, fire-resistant fluids, and synthetic fluids. Buna-N is the standard material used for piston seals and is suitable for most fluid power applications. Other materials, such as Viton or Polyurethane can also be used as seal materials depending upon the system temperature extremes and the type of fluid used in the system.

End Caps: An accumulator is provided with end caps at both ends to close the cylinder. The end caps are typically made from carbon steel with or without surface protection, stainless steel, or low-temperature carbon steel. An end cap contains an opening for the fluid port or gas port, as the case may be. Further, a piston accumulator with internal diameters up to 250 mm is fitted with a securing pin. This pin is intended to prevent the incorrect removal of the end cap.

Fluid Port, Piston Accumulator: The fluid chamber includes a standard fluid port with threads conforming to ISO (metric), DIN, ANSI (NPT), etc. or flanges conforming to DIN, ANSI, SAE, etc., for making a direct connection to the associated hydraulic system. The fluid port assembly is specially designed to prevent turbulent flow, excessive pressure drop, and potential pre-closure of the poppet valve.

Gas Valve, Piston Accumulator: A piston accumulator can be recharged the nitrogen gas with the help of a gas valve.

Sensors, Piston Accumulators: The location of the piston can be sensed or measured using a limit switch, proximity sensor, linear transducer, or mechanical indicating rod.

Safety Devices, Piston Accumulators: Many safety devices such as burst disc, gas safety valve, and temperature fuse plug are available for use in piston accumulators. For example, the burst disc is designed to rupture and hence to keep the gas port open, if the pressure on the gas side exceeds the maximum level. Further, a temperature fuse plug can be used to release the gas pressure completely when the temperature increases to an unacceptable level.

Installation, Piston Accumulators: Piston accumulators can operate in any mounting position. It may be noted that an accumulator installed at an angle or in a horizontal position will become a trap for fluid contaminants. Therefore, vertical installation is preferable for a piston accumulator, with the fluid side at the bottom of the accumulator. Moreover, vertical installation is essential for a piston accumulator with a position indicator. Large piston accumulators with a piston diameter greater than 350 mm must only be installed vertically.

Clamping Supports, Piston Accumulators: Clamping supports can be used preferably near the end caps to mount a piston accumulator for its proper support and isolation from system vibrations.

Back-up Gas Bottles, Piston Accumulators: A backup gas bottle is simply a gas cylinder connected to the charging valve assembly of an accumulator. Backup gas bottles are used in combination with a piston accumulator to provide additional gas volume for the accumulator without increasing the piston size. This allows for a large fluid delivery economically with only a small pressure differential. The accumulator operates as before, but because of the greater gas volume, the expansion is greater and therefore the fluid flows back into the system faster.

Advantages, Piston Accumulators: Piston accumulators are compact devices. They can provide higher flow rates (typically up to 250 l/s) than that of other types of gas-charged accumulators of comparable sizes. For higher flow capacity, several accumulators or gas bottles can be connected in parallel. Other advantages of piston accumulators are their portability and ability to handle a broad range of temperatures. A piston accumulator has a better damping capability as compared to other types of hydro-pneumatic accumulators due to hydraulic leakage as well as due to the friction between the piston and the shell.

Disadvantages, Piston Accumulators: The main disadvantages of piston accumulators are that they are susceptible to fluid contamination and they exhibit hysteresis due to seal friction. They are also expensive to manufacture and have practical size limitations. Typically, a gas-charged piston accumulator can cost twice as much as that of an equal-sized bladder accumulator. Further, it requires frequent pre-charging owing to increased gas leakage.

Moreover, it does not respond to transient pressures as fast as an equivalent bladder accumulator does, due to the greater mass of the piston. Also remember, a piston accumulator cannot be used as a shock absorber because of the inertia of the piston and the friction induced by piston seals.

Disassembly and Assembly Tips, Piston Accumulators

Only authorised persons, who are properly trained, should disassemble or assemble a piston accumulator. Next, the instructions in the operating manual should be followed. Remember, there is a danger to life due to parts of the accumulator flying off.

Before disassembling or assembling a piston accumulator, the system must always be depressurised. The fluid side should be relieved using a bleed valve and the gas in the gas chamber should be depressurised and the charge valve opened before the accumulator is disassembled. It should be ensured that the piston is moving freely, before removing the end caps. A rod can be used to check the free movement or otherwise jamming of the piston. The securing pin, if any, provided to lock the end cap must be taken out before removing the end cap.

Any welding, soldering or mechanical work should not be carried out on the piston accumulator, on any account.

Compliance with Rules and Regulations

All responsible owners and operators of hydraulic equipment with accumulators must adhere to the rules and regulations regarding the pressure vessels that are applicable at the place of installation of the equipment. It may be remembered that the owner is solely responsible for complying with existing provisions. General safety information for hydraulic accumulators is provided by relevant standards. Some of the safety standards of high-pressure vessels are given in Appendix 5.

Appendix 1 gives the specifications of piston accumulators.

Chapter 4 | Diaphragm Accumulators

A diaphragm (membrane) accumulator consists of two metallic hemispheres (shells), a flexible synthetic diaphragm, a fluid port, a gas valve, and mounting supports. The diaphragm is secured in between the two hemispheres. The diaphragm divides the inner hollow space of the accumulator into a fluid chamber and a gas chamber. A schematic diagram of a typical diaphragm accumulator is shown in Figure 4.1.

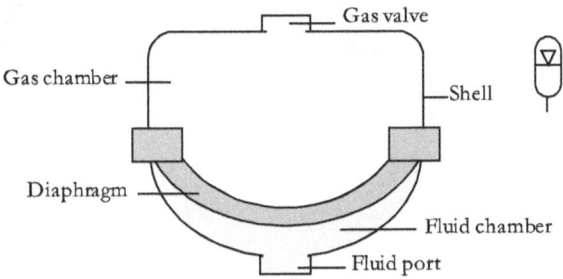

Figure 4.1 | A diaphragm type accumulator

Working, Diaphragm Accumulator: When the system pressure increases, the fluid is drawn into the fluid chamber, and the gas gets compressed. Any pressure drop in the system causes the diaphragm to expand, forcing out the stored fluid from the accumulator back into the system. That is, the compressed gas provides a quick pneumatic spring action to force stored energy from the accumulator into the system.

Shells, Diaphragm Accumulator: The shells in an accumulator can be welded or screwed to form the accumulator housing. A lock nut can be provided in a screw-type accumulator to hold the upper and lower sections of the accumulator tightly.

The shells are made of carbon steel, low-temperature steel, or high-tensile stainless steel. For use with a chemically aggressive fluid, the interior and sometimes the exterior of the carbon steel shells can optionally be provided with corrosion protection, with plastic coating using epoxy, or chemical plating using nickel or chrome. Alternatively, stainless steel can be used for the accumulator parts that are liable to be exposed to a corrosive environment or subjected to high pressure.

Diaphragm: The flexible diaphragm is a barrier that serves to divide the inner cavity of the accumulator into two chambers. One chamber is to contain hydraulic fluid on the system side and the other one is to hold nitrogen gas. The barrier must separate the hydraulic fluid chamber and the gas chamber in a leak-free manner. In the screw type accumulator, the diaphragm can be replaced. However, in the weld-type accumulator, it is not possible to replace the diaphragm.

Diaphragms materials are developed from the most advanced elastomers capable of meeting a wide range of temperatures from -40°C to +176°C. They must also be compatible with a wide variety of fluids including mineral-based fluids, fire-resistant fluids, and synthetic fluids.

Buna-N is the standard material used for the bladder and is suitable for most fluid power applications. Other materials, such as Viton (Fluorine rubber, FKM), Butyl, or Hydrin, can also be used as the diaphragm material for an accumulator to be used in a hydraulic system depending upon the system temperature limits and the type of fluid used in the system.

If an accumulator discharges rapidly with a high ratio of the maximum system pressure P_2 to the pre-charge pressure P_0 (P_2/P_0), the gas tends to cool below the permitted temperature. This cooling may cause cracking of the diaphragm material.

Valve Poppet, Diaphragm Accumulator: A valve poppet is set into the base of the diaphragm to prevent extrusion of the diaphragm into the opening for the fluid passage, in case the diaphragm over-expands.

Fluid Port, Diaphragm Accumulator: The fluid chamber includes a standard fluid port with threads conforming to ISO (metric), DIN, ANSI (NPT), etc. for making a direct connection to the associated hydraulic system. The fluid port assembly is specially designed to prevent turbulent flow, excessive pressure drop, and potential pre-closure of the poppet valve.

Gas Valve, Diaphragm Accumulator: In the weld-type and screw-type diaphragm accumulators, the gas can be supplied or discharged with the help of a normally-closed gas charging valve with a knob. Alternatively, in the weld-type accumulator, the gas chamber can be fully charged with nitrogen gas and then the opening passage used for the charging can be completely sealed. The gas chamber must be pre-charged to a definite pressure.

Mounting, Diaphragm Accumulators: By design, a diaphragm accumulator can be mounted in any position. However, in a system where contamination is a severe problem, a vertical mount with the fluid port of the accumulator oriented downward is, usually, preferred.

Support Clamps, Diaphragm Accumulators: Small accumulators up to a nominal volume of two litres can be screwed directly inline. An accumulator, used in an application where strong vibrations are expected, must be tightly secured using support clamps (fixing bands).

Key Features, Diaphragm Accumulators: The diaphragm accumulators are fast-acting and do not exhibit hysteresis. They are not affected by contamination, and they provide consistent behaviour under similar conditions. The speed of a diaphragm accumulator is governed by the gas, as there is no piston mass. Therefore, it reacts quickly to the changes in the system pressure. Hence, it is an excellent choice for an application requiring pressure pulsation damping.

Appendix 3 gives the specifications of diaphragm accumulators.

Chapter 5 | Bladder Accumulators

A bladder accumulator consists of a seamless cylindrical pressure vessel (shell), internal elastomeric bladder (bag), poppet valve, fluid port, charging valve, and clamps and brackets. A schematic diagram of a bladder (or bag) accumulator is shown in Figure 5.1.

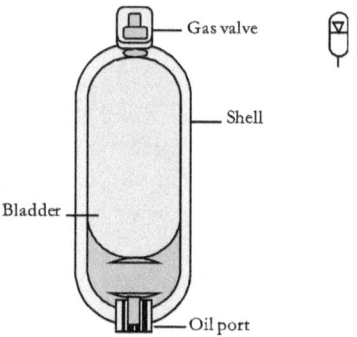

Figure 5.1 | A bladder accumulator

Working, Bladder Accumulator: When the system pressure increases, the fluid is drawn into the fluid chamber, and the gas gets compressed. Any drop in the system pressure causes the bladder to expand, forcing out the stored fluid from the accumulator back into the system.

Shell, Bladder Accumulator: The shell of a bladder accumulator is manufactured from homogeneous and seamless tubing that is, usually, heat-treated and stress relieved, as per the requirements of relevant standards, to ensure its excellent mechanical properties. Each end of the shell is formed in a hemispherical shape by spinning or forging.

The shell is made of carbon steel, high-tensile stainless steel, or low-temperature steel. For use with a chemically aggressive fluid, the interior and sometimes the exterior of the carbon steel shell can optionally be provided with corrosion protection, with plastic coating using epoxy or chemical plating using nickel or chrome. Alternatively, stainless steel can be used for the accumulator parts that are liable to be exposed to a corrosive environment or subjected to high pressure.

Bladder: Next, the bladder divides the shell into two chambers, namely, the fluid chamber on the system side and the gas chamber inside the bladder. The fluid chamber is to contain hydraulic fluid on the system side. The gas chamber inside the bladder is pre-charged to a certain pressure level. The bladder must separate the hydraulic fluid chamber and the gas chamber in a leak-free manner. A full range of bladders is developed from the most advanced elastomers capable of meeting a wide range of temperatures from -45°C to +150°C. They must also be compatible with a wide variety of fluids.

Buna-N is the standard material used for the bladder and is suitable for most fluid power applications. Other materials, such as low-temperature Buna-N (ECO), chloroprene, nitrile, Viton (Fluorine rubber, FKM), Butyl, ethylene propylene, and Hydrin, are also used as the bladder materials.

If an accumulator discharges rapidly with a high ratio of the maximum system pressure P_2 to the pre-charge pressure P_0 (P_2/P_0), the gas tends to cool below the permitted temperature. This cooling may cause cracking of the bladder material.

Poppet Valve, Bladder Accumulator: The fluid chamber in a bladder accumulator is provided with a spring-loaded poppet valve. This valve closes if the pressure on the gas side exceeds the pressure on the fluid side. In this way, the extrusion of the bladder into the downstream tubing is prevented, in case the bladder over-expands. If the minimum operating pressure is reached, about 10% of the nominal volume is to remain between the bladder and the poppet valve so that the bladder does not hit the valve in every expansion process.

Fluid Port, Bladder Accumulator: The fluid chamber includes standard ports with threads conforming to ISO (metric), DIN, ANSI (NPT), etc., or special bolt-on flanges conforming to DIN, ANSI, SAE, etc., for making a direct connection to the associated hydraulic system. The fluid port assembly is specially designed to prevent turbulent flow, excessive pressure drop, and potential pre-closure of the poppet valve.

Bottom-repairable and Top-repairable Bladder Accumulator Models: Bladder accumulators are designed as bottom-repairable and top-repairable models. In a bottom repairable model, the bladder can be inserted into the shell through the bottom opening in the shell, and the fluid port body and poppet valve assembly can be fixed to seal the accumulator. A top-repairable accumulator is designed as a cylindrical steel cylinder, open-top bladder, upper cap and threaded ring assembly, and fluid cap. The fluid cap with an opening is welded into the steel cylinder at the lower end. The open-top bladder can be inserted into the vessel and then can be retained and sealed by the upper cap and threaded ring assembly. A top-reparable model may be repaired easily without dismounting the accumulator. However, top-reparable models are comparatively expensive. Next, the bottom-repairable models are most popular.

Clamps and Brackets, Bladder Accumulators: Clamps and brackets (console) can be used to mount accumulators for their proper support and isolation from system vibrations. They are zinc-plated to resist corrosion and can be easily bolted or welded to hydraulic systems. A small accumulator can be fixed into its position using a light-duty clamp and a large accumulator can be mounted and supported using a heavy-duty clamp with a base bracket. Base brackets are used for the support of large vertically-mounted accumulators. The clamp is generally made of zinc-plated sheet steel or stainless steel strap. A rubber insert can be used in a clamp and a rubber support ring can be incorporated in a bracket to absorb any vibration and prevent noises from being transmitted through metal to metal contact.

Charging Kit, Bladder Accumulator: The gas chamber is pre-charged with nitrogen gas to a certain pressure level using a nitrogen source and a charging kit. The charging kit consists of a normally-closed charging valve with a knob, adapter, and pressure gauge. The adapter is used for the connection of the charging valve to the gas port. An additional valve can also be provided for the protection of the pressure gauge.

Installation, Bladder Accumulator: Bladder accumulators can be installed vertically, horizontally, or at any angle depending upon the application requirements. On a specific type of application, a particular position is preferable. For example, the vertical position of the accumulator is preferable for energy storage applications, and any installation position from vertical to horizontal is appropriate for pulsation damping applications.

Advantages and Limitation, Bladder Accumulator: Bladder accumulators are fast-acting and do not exhibit hysteresis. Therefore, a bladder accumulator is an excellent option for an application requiring pressure pulsation damping and shock suppression. It is not susceptible to contamination and provides consistent behaviour under similar conditions. The main limitation of the bladder accumulators is that they are larger than other types of comparable accumulators.

Features of Bladder Accumulators: Bladder accumulators are available as standard models or in custom-engineered designs. Standard models are available in sizes from 50 mm to 600 mm in diameter with fluid capacities from 0.06 to 1350 litres and operating pressures up to 1400 bar.

Due to the limited volume capacity of a bladder accumulator, banks of bladder accumulators can be connected to a manifold to provide the desired quantity of fluid to a system. However, this can cause physical space limitations in certain applications.

Appendix 2 gives the specifications of bladder accumulators.

Chapter 6 | Metal Bellows Accumulators

A metal bellows accumulator consists of a housing with metal bellows assembly. The metal bellows assembly divides the housing into a fluid section and a gas section. The bellows element is designed as either corrugated bellows or diaphragm bellows. The gas section is initially pre-charged and then permanently sealed. The working of a metal bellows accumulator is similar to that of piston accumulators. Figure 6.1 shows the schematic diagram of a metal bellows accumulator.

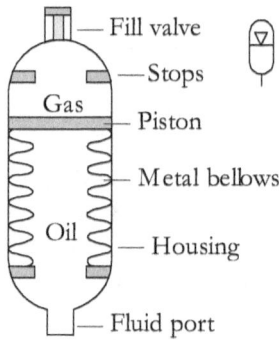

Figure 6.1 | A schematic diagram of a metal bellows accumulator.

They are very reliable and usually, have a long service life. Moreover, they are gas-tight, maintenance-free, and media resistant over a broad range of temperatures.

However, a metal bellows accumulator responds slowly to any pressure changes due to the increased mass of its piston and bellows.

They are used where a fast response time is not critical, yet reliability is essential. They are used in a wide range of applications, including aircraft systems, missile systems, ships, military ground vehicles, and chemical industries.

Chapter 7 | Comparison of Accumulators

Table 7.1 gives a comparison of diaphragm, bladder and piston type hydraulic accumulators. The values given are typical.

Table 7.1 | Comparison of accumulators

Parameter / property	Diaphragm	Bladder	Piston
Size	≤3.5 l	≤54 l	≤1350 l
Working pressure	250 bar	690 bar	2500 bar
Flow rate	≤150 lpm	≤900 lpm	≤900 lpm
Compression ratio	8:1	4:1	10:1
Flow rate (Typical)	≤1 litre/s	≤15 litres/s	≤215 litres/s
Seal materials	Buna-N Viton Butyl Hydrin	Buna-N Butyl Viton Hydrin	Buna-N Viton Polyurethane
Temperature limits	-40°C to 176°C	-45°C to 150°C	-40°C to 150°C
Heaviness	Light-weight	Medium-weight	Heavy
Cost	Low	Medium	High
Application	Suitable for small volume and flow rates	Best for general purpose applications	Best for large volumes or high flow rates
Shock suppression	Good	Good	Not good
Mounting position	Any position	Any position	Only vertical
Fluid port connection	SAE, NPT	SAE, NPT	SAE, NPT

Chapter 8 | Pre-charging of Accumulators

A hydro-pneumatic accumulator must be filled with dry inert gas, such as nitrogen gas, while there is no fluid in the fluid chamber. The pre-charge level of the gas medium is an essential parameter for the gas accumulator since the pre-charge pressure along with the accumulator volume determines the maximum amount of hydraulic energy that can be stored in it.

Each hydraulic application has its pre-charge requirements. In general, a gas accumulator is pre-charged to a certain percentage of the minimum system pressure, [For a definition, see Page 40, Chapter 14] depending upon the type of accumulator and application, and as per the recommendation of its manufacturer. Typically, the pre-charge pressure for energy storage applications can be taken as 80 to 90% of the minimum system pressure. The pre-charge pressure for a pulsation compensator or shock absorber can be taken as 65 to 80% of the minimum system pressure.

Diaphragm accumulators are generally delivered without pre-charge pressure. Bladder accumulators are generally delivered with a nitrogen pre-charge pressure of approximately 2 to 5 bar. After installation and before initial start-up, the pre-charge pressure must be set to the requirements of the application or as per the machine manufacturer's specifications.

Checking of Pre-charge Pressure
The pre-charge pressure of an accumulator should be checked at regular intervals. Typically, the first check should be carried out after one week of its operation, the second check after three months, the third check after one year, and thereafter continue once every year.

If the pre-charge is low, investigate the cause and rectify it. The possible causes for the low pre-charge pressure include gas leakage, damaged gas valve, or damaged bladder/diaphragm.

Chapter 9 | Safety Requirements of Accumulators

A hydraulic accumulator is a pressure vessel that stores an enormous amount of potential energy for the subsequent release of the same for performing some useful hydraulic functions. Accumulators can be dangerous to personnel and property if they discharge the stored pressure inadvertently. Moreover, they are subject to regulations applicable to the place of their installation.

Therefore, it is necessary to isolate the accumulators from the associated systems and discharge the pressures from the accumulators, during periods of maintenance or emergency. Typically, safety devices must be incorporated in an accumulator to provide a shut-off facility and pressure-limiting and pressure relief features. So, it is recommended to always use a safety-and-shut-off block along with an accumulator for protecting the system and the personnel against hazardous stored energy. Figure 9.1 shows the circuit diagram of a safety-and-shut-off block.

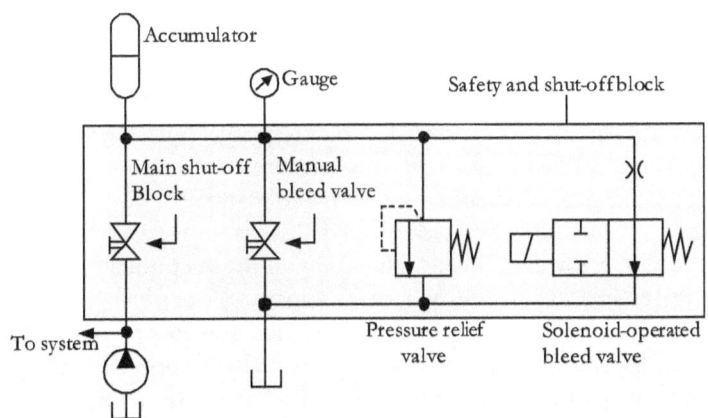

Figure 9.1 | A circuit layout of a safety-and-shut-off block for a hydraulic accumulator

Safety-and-Shut-off Block

Figure 9.2 shows the schematic diagram of a safety-and-shut-off block connected to an accumulator. It is a multifunctional valve system connected to a hydraulic accumulator. Further, it consists of a shut-off valve, a manual bleed valve, a pressure relief valve, an optional 2-way solenoid-operated bleed valve and a pressure gauge. It is possible to configure the block in a modular fashion with a host of connection options, and this option permits its versatility for mounting.

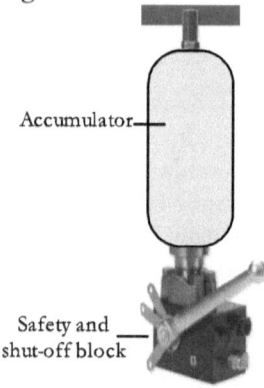

Accumulator

Safety and
shut-off block

Figure 9.2 | A schematic of a safety-and-shut-off block

The shut-off valve is used to isolate the accumulator instantly from the hydraulic system, for maintenance purpose or in an emergency. Once isolated, the accumulator can be safely discharged to the reservoir through the manual bleed valve. The optional solenoid-operated bleed valve, if used, assists in the automatic release of the stored energy in the accumulator in case of an emergency shutdown or the loss of electrical power. A pressure relief valve protects the accumulator from over-pressurisation. The safety and shut-off block should always be mounted close to the accumulator. Remember, the commissioning and maintenance of the safety block and associated equipment must be performed only by qualified technical staff. Specification parameters of accumulators and safety blocks are given in Appendix 4.

Chapter 10 | Applications of Accumulators

Accumulators are extensively used in industrial and mobile applications for the suppression of shocks, storage of energy, compensation of leakages, and energy recovery.

Accumulators are found on industrial machinery, mobile equipment, and in marine, oil and gas, and aerospace applications. For example, they are used in industrial applications such as machine tools, steel production, metal-forming machinery, foundries, paper production, power transmission, injection moulding, and die casting, and mobile applications such as mining, construction, forestry, and agriculture.

For shock suppression, they are used in large hydraulic presses, vehicles, and equipment used in the construction, offshore, and mining fields.

An accumulator can supplement the pump flow in a hydraulic application, such as an aircraft landing gear, which requires a considerable volume of fluid.

The use of accumulators as an energy storage device is convenient in operation such as the emergency completion of a working cycle in case of the failure of the primary power source.

The inclusion of accumulators in applications permits the use of small pumping stations in applications, such as cranes, dock gates, lifts, and forging presses, which do not use the entire pump flow for most of the machine cycle time.

The weight-reduced diaphragm, bladder, and piston accumulators find applications in the aircraft industry, offshore industry, city buses, garbage trucks, wind power industry, automotive industry and railway vehicles for energy storage and reducing energy consumption.

Chapter 11 | Basic Accumulator Circuits

Hydraulic accumulators perform many functions in hydraulic systems. They are basically employed for shock absorption and energy storage. The following sections explain many typical hydraulic circuits for realising these functions.

An accumulator as Hydraulic Shock Absorber
Figure 11.1(a) shows a simple hydraulic circuit used for the direction control of a double-acting cylinder using a 4/3–way, closed-centre valve.

As we are aware, momentary high-pressure surges or shock waves are likely to be generated in the system while shifting the valve rapidly. These shock pressures are developed because of the failure of the associated PRV to act fast enough to drain off the high-pressure fluid from the circuit. These high-pressure surges can be dangerous to personnel and equipment.

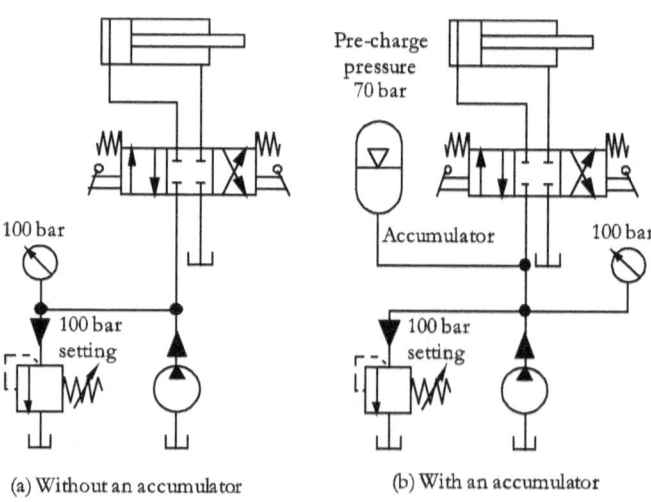

(a) Without an accumulator   (b) With an accumulator

Figure 11.1 | Basic hydraulic circuits

A correctly-sized accumulator can be connected to the circuit, as shown in Figure 11.1(b), to suppress these shock pressures. The accumulator absorbs the shock pressures, whenever they appear in the circuit. However, it can be observed that this circuit is not capable of storing the excess energy that has gone into the accumulator, as the pressure relief valve in the circuit slowly drains the excess fluid back to the system reservoir.

Other functions, such as storing energy, unloading the pump upon reaching the system pressure, and setting the discharge flow rate from the accumulator to suit the system requirement, can be achieved by using additional components. The following sections explain the circuits for realising these accumulator functions.

An accumulator as an Auxiliary Power Source

Figure 11.2 gives different positions of a hydraulic circuit necessarily with a fixed-displacement pump, reservoir, unloading valve, and accumulator. The accumulator should be sized appropriately to store the energy during periods of low demand and to use the stored energy as an auxiliary source of power during periods of high demand.

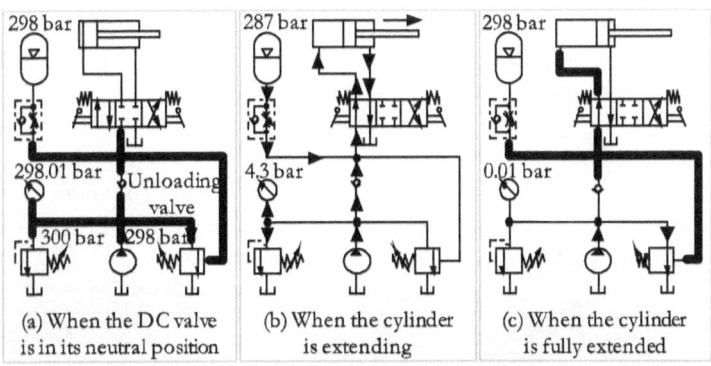

| (a) When the DC valve is in its neutral position | (b) When the cylinder is extending | (c) When the cylinder is fully extended |

Figure 11.2 | Three positions of a circuit with an accumulator acting as an auxiliary power source

Figure 11.2(a) shows the position of the circuit when the 4/3-way, closed-centre, valve is pulled to its neutral position. In this position, the pump charges and builds up pressure in the accumulator through the check valve. The unloading valve remains closed until the set pressure is reached. The unloading valve opens and unloads the pump flow back to the reservoir at low pressure when the pressure setting of the valve is reached. The unloading of pump flow allows the pump to operate at a minimal load. The isolation check valve prevents the accumulator fluid from flowing back to the pump and traps the pressurised fluid in the accumulator.

Figure 11.2(b) shows the position when the 4/3-way valve is shifted to its left envelope for the forward stroke of the cylinder. During this position, the pump flow is supplemented by the accumulator by releasing the stored energy. Remember that there is a volume of fluid at an elevated pressure in the accumulator that can discharge itself almost instantaneously into the system. Therefore, when the accumulator is releasing the stored energy, it is necessary to use a throttle valve to set the flow rate well within the requirement of the system.

Figure 11.2(c) shows the position of the circuit when the cylinder is fully extended, and the accumulator is fully charged. When the set pressure is reached, the unloading valve opens and unloads the pump to the reservoir at low pressure.

When the 4/3-way DC valve is shifted to its right envelope for the return stroke of the cylinder, similar kinds of actions, as explained in the previous paragraphs, can be expected from the accumulator and the unloading valve.

Accumulator Circuit with an Unloading and Dump valve

It is necessary to discharge the pressurised fluid in an accumulator either manually or automatically when the system pump stops, as a safety precaution. A solenoid-operated or a pilot-operated dump valve can be used to discharge the stored fluid automatically when the pump stops. The dump valve is designed to close when the pump runs and open when the pump stops. Parts (a) and (b) of Figure 11.3 show different positions of the hydraulic circuit with a fixed-volume pump, reservoir, accumulator, check valve, unloading valve, and 2/2-way pilot-operated dump valve.

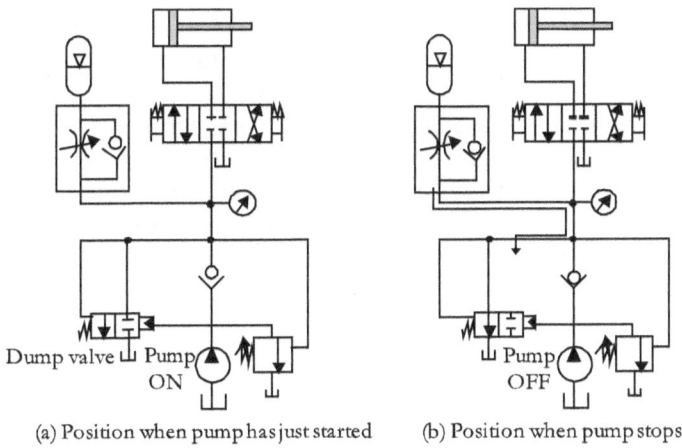

(a) Position when pump has just started   (b) Position when pump stops

Figure 11.3 | Two positions of an accumulator circuit with an automatic pilot-operated dump valve

Figure 11.3(a) shows the position of the circuit as soon as the pump is switched on. The flow is directed to the accumulator through the opened check valve. The dump valve closes immediately and remains closed when the accumulator keeps charging. When the set pressure of the unloading valve is reached, the valve opens and unloads the pump flow to the reservoir at low pressure. It may be noted that the dump valve remains closed, even when the unloading valve is relieving the excess pressure.

Figure 11.3(b) shows the position of the circuit, as soon as the pump is switched off. Here, the pilot pressure to the dump valve drops, allowing its spool to shift position. The dump valve now opens, and the pressurised fluid in the accumulator has a path directly to the reservoir through the dump valve. Therefore, the accumulator can discharge quickly and automatically when the pump is turned off, making it safe to carry out any maintenance work on the system.

A pilot-operated dump valve works well in most cases, but it can cause problems in some situations. For example, if the dump valve fails to open when the pump stops, the circuit is unsafe. This possibility is a safety hazard to maintenance personnel. Therefore, it is advisable to be cautious and check an accumulator circuit for the trapped pressure before working on it, even if an automatic unloading arrangement is provided.

Chapter 12 | Maintenance of Accumulators

Observing safe practices while working with accumulators and carrying out their proper maintenance are the most critical activities while dealing with them, for ensuring the safety of the associated equipment and personnel. Maintenance personnel should know the rules related to the pressure vessels, such as accumulators. It is also essential to take all safety precautions in hydraulic accumulators against the hazardous stored energy with built-in PRVs, shut-off valves, and solenoid and/or manually operated bleed valves.

Therefore, an accumulator used in a hydraulic system should be designed to shut off the accumulator, discharge trapped fluid, and protect the system, in an emergency or during a system shutdown. The solenoid-operated bleed valve in the accumulator allows the automatic release of the energy trapped in the accumulator in an emergency or during the shutdown of the associated system. The following sections present the general maintenance guidelines of hydraulic accumulators, the details of their installation and the pre-charging of the gas-loaded accumulators.

General Guidelines for the Maintenance of Accumulators

The following bulleted lines enlist some necessary maintenance and safety guidelines for the hydraulic circuit with an accumulator. These guidelines are most general, and they are not intended for any specific hydraulic machine.

- Only qualified maintenance technicians must carry out the maintenance work on the system with the accumulator.
- Attach warning sign such as **'ATTENTION: System with Accumulator'** close to the accumulator.
- The accumulator should be provided with the safety valve block with a PRV, shut-off valve, and bleed valve.

- Always depressurise and isolate the accumulator before servicing the system.
- Never start the system before charging the accumulator.
- Always maintain the maximum working pressure, pre-charge pressure, and operating temperature of the accumulator within acceptable limits.
- If the accumulator is a gas-charged one, make sure that the charging and discharging rates of the accumulator should be restricted to reasonable values to avoid damage to the accumulator and other system components.
- Ensure that the accumulator is manufactured, tested, and certified as per the statutory standards, like ASME.
- Do not operate a system with an accumulator circuit until the accumulator is securely anchored to a stable structure.

Accumulator Installation
An accumulator in the hydraulic circuit should be installed very close to the source of shock or potential energy, to reduce the pressure loss on the line between them. It must be correctly installed in an easily accessible place using robust collars. The markings engraved on the accumulator must remain visible after its installation. Typically an accumulator is installed in a vertical position with the fluid connection port in its bottom part. However, if the accumulator has to be mounted horizontally due to space constraints, a loss of efficiency, as well as a reduction in its service life, are to be expected. Further, the horizontal arrangement could become a trap for the contaminants.

Inspection of Accumulators
Periodic inspection of especially a large-volume or high-pressure accumulator is required to find any developing symptoms of a potential malfunction of the accumulator. The inspection may be carried out at predetermined intervals such as every year or as per the instructions of the manufacturer. Remember, the owner and the operator of the equipment are responsible for ensuring the safe operation, and thoroughness and frequency of the inspection of the pressure vessel.

Chapter 13 | Pre-charging Procedure

A gas-charged accumulator is delivered with a bare minimum pre-charge pressure. The accumulator can be charged before or after it has been installed on the system. Before charging a bladder accumulator, it is advisable to pour some fluid into the accumulator to allow the fluid to coat the inside of its shell. This fluid layer provides the initial lubrication between the bladder and the shell.

The usual pre-charge pressure for an accumulator is in the range of 80% to 90% of the minimum system pressure. For a hydraulic system where the accumulator is being utilised as a shock absorber or pulsation dampener, a pre-charge pressure of 65% of the minimum system pressure can be used. Also, the percentage of the pre-charge pressure should not fall below a specific value. If the pre-charge pressure in the accumulator goes beyond its maximum or minimum limits, its elastomeric part is liable to be damaged.

If the nitrogen gas is allowed to flow too rapidly into the accumulator, it can lead to the chilling of the polymeric material of the accumulator's diaphragm or bladder. This chilling effect may result in the immediate brittle failure of the polymeric material.

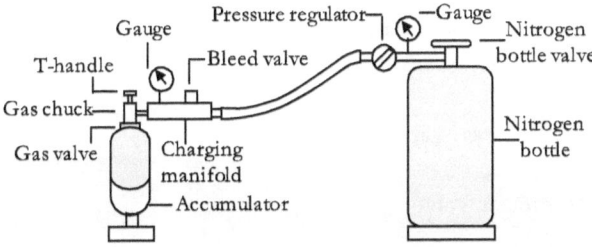

Figure 13.1 | A schematic diagram showing the set-up for pre-charging a hydraulic accumulator

Initially, an accumulator should be pre-charged with the clean and dry nitrogen gas of class 4.0 purest (N2 content 99.99% by volume) using a charging kit. Figure 13.1 shows the schematic diagram indicating the arrangement for pre-charging the accumulator. This package mainly consists of a bottle with nitrogen gas, a charging manifold, a gas chuck, and connecting hoses. The manifold consists of a bleed valve and a pressure gauge. The nitrogen bottle has a gas valve and pressure regulator. The accumulator is, usually provided with a gas valve and protective cap. The following bulleted lines highlight the procedure for the pre-charging of a hydraulic accumulator.

- Attach the charging kit hose to the gas chuck on one side and the regulator on the other side

- Close the bleed valve on the charging manifold

- Attach the chuck to the accumulator gas valve

- Open the gas valve by turning the chuck's T-handle

- Open the nitrogen bottle valve slowly and fill the accumulator to the desired pre-charge pressure

- Close the nitrogen bottle valve

- If the desired pre-charge is exceeded, open the bleed valve, to relieve the excess pressure

- Close the bleed valve

- Close the gas valve

- Remove the gas chuck

Chapter 14 | Accumulator Sizing

Gas-charged accumulators are widely used in many industries to store energy for intermittent duty cycles or to provide a source of standby power. They are typically rated in terms of the gas volume they can have when all the fluid has been discharged. A hydro-pneumatic accumulator must be sized optimally and pre-charged to the correct pressure to meet the assigned function.

The accumulator sizing is affected by the compression (or expansion) process of the gas. If the charging (or discharging) of the accumulator takes place so slowly, there is sufficient time for heat to be added (or subtracted) by the accumulator wall to maintain a constant gas temperature (isothermal process). If the charging (or discharging) takes place rapidly, there is not enough time for sufficient heat transfer through the accumulator walls (adiabatic process). These conditions are theoretical, and the actual charging (or discharging) process takes place between these two ideal conditions. The actual charging (or discharging) process is called the polytropic process.

It is, however, possible to state, with reasonable accuracy, that when an accumulator is used as a volume compensator or leakage compensator, the condition is isothermal. On the other hand, when an accumulator is used in other applications, such as energy storage, pulsation damper, emergency power source, or shock absorber, the condition is close to adiabatic.

To calculate the capacity of the accumulator for the system requires intimate knowledge of the operating conditions of the system. Then, using any one of the mathematical models, as explained below, the size of the accumulator can be calculated. Manufacturers also bring out sophisticated software packages for the accurate determination of the capacity of an accumulator.

Specific definitions would be useful for further explanation:

- **Maximum System Pressure (P_2):** This is the maximum (nominal) pressure developed in the system, as per the setting of the system relief valve. It is usually the no flow rating of the hydraulic pump.

- **Minimum System Pressure (P_1):** This is the minimum pressure that the accumulator must maintain in the hydraulic system. This pressure is a design requirement used to size the accumulator.

- **Pre-charge Pressure (P_0):** It is the pressure of the nitrogen in an accumulator without any hydraulic fluid in the accumulator. The pre-charge pressure determines the amount of fluid that an accumulator can hold at the system pressure and the desired minimum hydraulic system pressure.

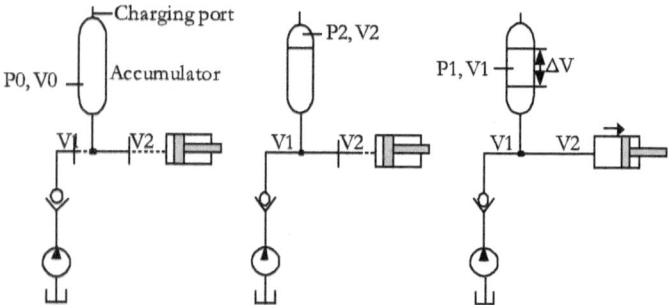

(a) Pre-charge position (b) Fully-charged position (c) Minimum charge position

Figure 14.1 | A schematic diagram showing three positions of a gas-charged accumulator connected in a hydraulic system

Figure 14.1 shows different positions of a hydraulic system with a pump, cylinder, check valve, and gas-charged accumulator for supplementing the power. Directional control (DC) valves V1 and V2 are also used in the circuit. Let D be the bore diameter

40

and S be the stroke length of the cylinder. Figure 14.1(a) shows the position of the accumulator when it is almost pre-charged. Let P_0 [= (0.95 to 0.97) x P_1 (i.e., Minimum system pressure)] be the pre-charge pressure in the accumulator and V_0 be its corresponding volume. V_0 is the required accumulator size.

Figure 14.1(b) shows the position of the hydraulic circuit when the pump is turned on keeping the DC valve V_1 open, and the valve V_2 closed. During this period, the accumulator gets charged to the maximum pressure corresponding to the setting of the pressure relief valve (not shown) in the system. Let P_2 be the maximum charge pressure in the accumulator and V_2 be the corresponding volume during this stage. Figure 14.1(c) shows the position of the circuit when both the DC valves V_1 and V_2 are open, and the cylinder has just reached the end of its stroke. During this period, the accumulator discharges fluid into the system, and the pressure in the accumulator decreases. Let P_1 be the pressure in the accumulator during this stage and V_1 be the corresponding volume. It may be noted that the fluid capacity available in the accumulator to drive the cylinder is $(V_1 - V_2)$.

In the isothermal process, the compression and expansion of the gas in an accumulator take place slowly, so that the existence of an approximately constant temperature condition can be presumed, as the complete heat transfer is possible between the gas medium and the environment. In the adiabatic process, the compression and expansion of the gas are quick so that no heat transfer is possible between the gas medium and the environment. The required accumulator size for various charging and discharging conditions can be derived in the following ways.

Charging and Discharging under Isothermal Condition
The perfect gas laws govern the compression and decompression of the nitrogen gas contained in the accumulator. Using the Boyle-Mariotte's law for ideal gases, assuming slow charging and slow discharging, to allow the gas in the accumulator to maintain

its temperature close to a constant, we get the size of accumulator V_0 from the following equations:

$P_0 V_0 = P_1 V_1 = P_2 V_2$
That is, $V_1 - V_2 = [(P_0 V_0) / P_1] - [(P_0 V_0) / P_2]$
$= V_0 [(P_0 / P_1) - (P_0 / P_2)]$
Therefore,
Accumulator volume, $V_0 = [V_1 - V_2] / [(P_0 / P_1) - (P_0 / P_2)]$

$(V_1 - V_2)$ is the fluid volume to be supplied by the accumulator for the extension stroke (x% of the full cylinder volume).

That is, $V_1 - V_2 = (x \%) \times (\prod D^2 / 4) \times S$

As the gas is pressurised, its temperature is liable to go up, and the volume of the fluid entering the accumulator is lower than the calculated amount. This limitation can be compensated by increasing the accumulator capacity by about 5%. If there are many actuators in the system, consider the peak loading when sizing the accumulator.

Charging and Discharging under Adiabatic Condition
Most fluid power designers use the ideal gas laws for the design calculations of accumulators. However, the primary gas laws do not apply when there is no or little heat transfer into or out of an accumulator, as found in a shock absorber or pulsation damper or emergency power source. Remember, today's hydraulic systems move faster with higher cycle rates. There exists a short time for heat to enter or leave the accumulator, so we assume that the compression and expansion of the gas be adiabatic – that is; no heat is transferred into or out of the accumulator. Assuming quick charging and quick discharging, we get the size of accumulator V_0 from the following equation:

$$P_0 V_0^n = P_1 V_1^n = P_2 V_2^n$$

Where n is the polytropic exponent (n= 1.4 for diatomic gas)

The P-V diagram of Figure 14.2 gives the pressure-volume relationship inside a gas-charged accumulator.

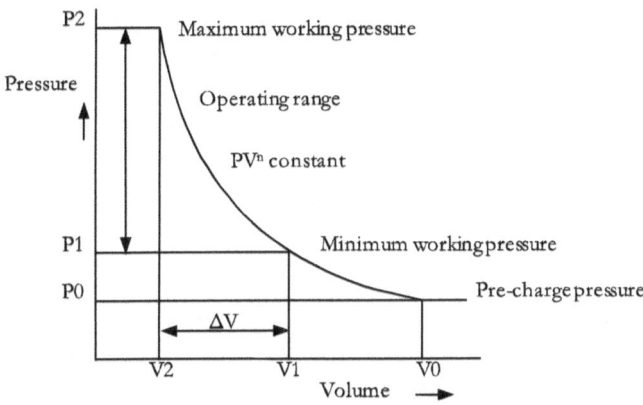

Figure 14.2 | PV diagram of a gas-charged accumulator

Similar to what is shown for finding the size of the accumulator with isothermal charging and discharging process, we have the size of the accumulator with charging and discharging under the adiabatic condition as:

Accumulator volume, $V_{0, \text{adiabatic}}$
$$= [V_1 - V_2] / [(P_0 / P_1)^{1/n} - (P_0 / P_2)^{1/n}]$$

Note:
1. Use the ratio of absolute pressures, not the gauge values.
2. The accumulator calculations are influenced by both the operating temperatures and pressures.

Slow Charging and Quick Discharging
Accumulator capacity when the charging process is slow (isothermal), and the discharging process is quick (adiabatic) can be calculated from the following equation:

Accumulator volume, V_0
$$= [V_1 - V_2] / \{(P_0 / P_2)^{1/n} - [(P_2 / P_1)^{1/n} - 1]\}$$

Temperature Influence, Accumulator Sizing

The operating temperature is liable to be changed considerably during the charging and the discharging cycle of an accumulator. This variation in the temperature should be taken into consideration when the accumulator volume is calculated. Let the capacity of the accumulator be V_0 at the temperature T_1 (K), and the capacity of the accumulator has increased to V_{oT} when the temperature has risen to T_2 (K). The capacity (V_{oT}) can be calculated by using the following formula:

Accumulator volume, $V_{oT} = V_0 \times (T_2/T_1)$

Correction Coefficient at Higher Pressures

The nitrogen gas in accumulators does not behave according to the ideal gas laws when the pressure goes beyond 200 bar. The capacity of the accumulator (V_{oP}), at higher pressures, under isothermal and adiabatic conditions, can be calculated by using the following formula:

Accumulator volume, $V_{oP} = V_0 / C_i$ (for isothermal condition)
$$= V_0 / C_a \text{ (for adiabatic condition)}$$

Where C_i is the isothermal correction coefficient, and C_a is the adiabatic correction coefficient, both have to be deduced from the standard charts published by manufacturers.

Useful Tips, Sizing of Hydro-pneumatic Accumulators

1. To achieve the best utilization of the accumulator volume possible as well as long service life, the pre-charge pressure at the maximum operating temperature should be about 90% of the maximum system pressure.

2. The maximum system pressure (P_2) is not to exceed four times the pre-charge pressure (P_0) [$P_2 \leq 4 \times P_0$] to prevent excess strain on the elasticity of the bladder.

3. The smaller the difference between P1 and P2, the longer will be the service life of the bladder. However, the degree of utilization of the maximum storage capacity will be reduced.

Example 14.1 | A moulding press has to be kept closed during its curing phase of 90 minutes at a constant pressure of 190 bar. After the mould has been closed, the pump is switched off. The expected fluid leakage is at the rate of 4 cm³/min, and the minimum permissible pressure is 188 bar, during the curing period. What is the capacity of a gas-charged accumulator used to compensate for the leakage? Assume the pre-charge pressure as 178 bar.

Solution

Leakage rate   = 4 cm³/min

Curing period   = 90 min

Working pressure, P_2 = 190 bar = 191 bar (a)

Minimum permissible pressure, P_1
$$= 188 \text{ bar} = 189 \text{ bar (a)}$$

Pre-charge pressure, P_0 = 178 bar = 179 bar (a)

Let V_0, V_1, and V_2 be the accumulator volumes during its pre-charge, fully charged and minimum charge positions respectively.

Total leakage during the curing period = 0.004x90 = 0.36 litre

Therefore, $V_1 - V_2$ = 0.36 litre

Capacity of the accumulator, V_0
$$= (V_1 - V_2)/ [(P_0 /P_1) - (P_0 /P_2)]$$
$$= (0.36)/ [(179 /189) - (179/191)] = 36.3 \text{ litres}$$

A standard capacity accumulator closest to the calculated value must be selected.

Example 14.2 | A gas-charged accumulator used as an emergency power source charges and discharges quickly at the rate of 2 litres in 2 seconds. It has a minimum working pressure of 70 bar and a maximum working pressure of 135 bar, and the operating temperature range is from 20 °C to 80 °C. Assume the pre-charge pressure as 95% of the minimum pressure. Calculate the accumulator capacity.

Solution

Charging/discharging rate   = 2 litres in 2 seconds
Minimum working pressure, P_1 = 70 bar = 71 bar (a)
Minimum working pressure, P_2 = 135 bar = 136 bar (a)
Minimum temperature, T_1 = 20 °C = 20+273 K = 293K
Maximum temperature, T_2 = 80 °C = 80+273 K = 353K

Assume the value of the polytropic exponent as 1.4.

Pre-charge pressure, P_0
  = 0.95xP1=0.95x70 bar = 66.5 bar = 67.5 bar(a)

As the accumulator is used as an emergency power source, the charging and discharging of the accumulator is quick, and the charging and discharging operations can be considered as adiabatic. Therefore,

Accumulator capacity, adiabatic, V_0
  $= \Delta V / [(P_0 / P_1)^{1/n} - (P_0 / P_2)^{1/n}]$
  $= 2 / \{(67.5 / 71)^{1/1.4} - (67.5 / 136)^{1/1.4}\}$ =5.58 litres

Considering the temperature variations,

Accumulator capacity, V_{oT} $= V_0 \times (T_2/T_1)$
  $= 5.58 \times (353/293) = 6.72$ litres

A standard capacity accumulator closest to the calculated value must be selected.

Example 14.3 | A gas-charged accumulator must discharge 4.6 litres of fluid for 3 seconds with a change of pressure from 280 bar to 208 bar [3017 psi]. The loading time is 4 minutes. Define the capacity keeping in mind that the ambient temperature changes from 20°C to 50°C. Assume the adiabatic and isothermal correction factors for higher pressures C_a as 0.72 and C_i as 0.83respectively.

Solution

Discharge rate = 4.6 litres in 3 s
Minimum operating pressure, P_1 = 208 bar = 209 bar (a)
Maximum operating pressure, P_2 = 280 bar = 281 bar (a)
Minimum temperature, T_1 = 20 ^{0}C = 273+20 K = 293 K
Maximum temperature, T_2 = 50 ^{0}C = 273+50 K = 323 K
Correction coefficient, adiabatic, C_a = 0.72
Correction coefficient, isothermal, C_i = 0.83
Assume the value of the polytropic exponent as 1.4.

Pre-charge pressure, P_0 = 0.95x208 = 198 bar = 199 bar(a)
Correction coefficient, C_{av} = (0.72 + 0.82)/2 = 0.77

As the accumulator storage is slow, (isothermal), and discharge is quick (adiabatic), accumulator capacity can be calculated from the following equation:

Accumulator capacity, V_0
= $[V_1 - V_2] / \{(P_0 /P_2)^{1/n} \times [(P_2 /P_1)^{1/n}-1]\}$
=4.6 / $\{(199 /281)^{1/1.4} \times [(281 /209)^{1/1.4}-1]\}$ = 25 litres

Now, considering the correction coefficient for high pressure and the temperature change, we have:
Accumulator capacity, V_{oT} = $(V_0/C_{av}) \times (T_2/T_1)$
= (25/0.77) x (323/293) = 35.79 litres

An accumulator closest to the calculated value must be selected.

15 | Objective Type Questions

1. A function of hydraulic accumulators is:
    a) Limiting pressure
    b) Dampening pulsations
    c) Controlling speed
    d) Conditioning fluid

2. Mark, the <u>incorrect</u> statement, about hydraulic accumulators
    a) Piston accumulators have a higher compression ratio.
    b) Piston accumulators can be used as pulsation dampener or shock absorber.
    c) A diaphragm accumulator is an excellent choice for applications requiring pressure pulsation damping.
    d) The diaphragm accumulators are fast-acting and do not exhibit hysteresis.

3. The best type of hydraulic accumulator for large volumes or high flow rates is:
    a) Diaphragm
    b) Bladder
    c) Piston
    d) Spring-loaded

4. The gas used in a hydro-pneumatic accumulator is:
    a) Air
    b) Nitrogen
    c) Oxygen
    d) All of the above

5. Proper maintenance of gas-charged accumulators involves:
    a) Shutting off the accumulator before maintenance
    b) Discharging trapped fluid
    c) Frequent replacement of nitrogen gas
    d) Both (a) and (b)

16 | Review Questions

1. Define a hydraulic accumulator.
2. Explain the essential operation of a hydraulic accumulator using a simple circuit.
3. List four essential functions of hydraulic accumulators.
4. Explain how the pressure pulsations in a hydraulic rock drilling operation can be suppressed.
5. Describe the function of a hydraulic accumulator used as a shock absorber.
6. What adverse effect can happen in a primary hydraulic loader system with a 2-ton front-loading bucket is suddenly stopped? How can this effect be overcome?
7. Explain the function of a hydraulic accumulator used as a pump supplement.
8. Explain how the capacity requirements of the pump, utilised in a hydraulic system delivering only a short period of high-volume flow during a given cycle, can be reduced.
9. Develop a hydraulic circuit for carrying out a work process where significant fluid flows for a short period in each cycle.
10. Write briefly about the three most common applications of hydraulic accumulators?
11. Classify hydraulic accumulators.
12. Briefly explain about three basic types of accumulators used in hydraulic systems?
13. Draw the symbols for the weight-loaded, spring-loaded and gas-charged hydraulic accumulators.
14. Explain the principle of operation of hydraulic accumulators.
15. Briefly explain the operation of a weight-loaded hydraulic accumulator, with a simple sketch.
16. Give some advantages and disadvantages of the weight-loaded hydraulic accumulators.
17. Briefly explain the operation of a spring-loaded hydraulic accumulator, with a simple sketch.
18. Give some benefits and disadvantages of the spring-loaded hydraulic accumulators.

19. Name three major classifications of the gas-charged hydraulic accumulators. Give one advantage of each accumulator.
20. Explain the fundamental operating principle of hydropneumatic accumulators.
21. What are the advantages of the gas-charged accumulators as compared to other types of hydraulic accumulators?
22. Why is the non-separator-type gas-charged accumulator not used in modern hydraulic systems?
23. Briefly explain the operation of a piston-type hydraulic accumulator, with a simple sketch.
24. Briefly explain the constructional features of piston-type hydraulic accumulators
25. Give one advantage, and one disadvantage of the piston type hydraulic accumulators.
26. Briefly explain the principle of operation of a hydraulic diaphragm accumulator, with a simple sketch.
27. Briefly describe the constructional features of the hydraulic diaphragm type accumulators
28. Give one advantage and one disadvantage of the hydraulic diaphragm accumulators.
29. Briefly explain the operation of a hydraulic bladder accumulator, with a simple sketch.
30. Explain the function and purpose of a gas-filled accumulator.
31. Briefly explain the constructional features of the hydraulic bladder type accumulators
32. Give one advantage and one disadvantage of the hydraulic bladder accumulator.
33. Compare the diaphragm, bladder and piston accumulators for typical sizes, pressures, flow rates and suitability for shock suppression.
34. Give Short notes on (a) pre-charging the gas-loaded hydraulic accumulators and (b) the use of dump valves in hydraulic accumulator circuits.
35. What is the reason for connecting a throttle check valve at the inlet/outlet port of some accumulators?

36. Explain how a load-induced shock can be reduced in a hydraulic system, with a circuit diagram.
37. Explain the safety requirements of accumulators in hydraulic circuits.
38. Describe the functions of the safety-and-shut-off block associated with hydraulic accumulators.
39. Explain why it is required to unload an accumulator automatically when the machine is shut down.
40. Draw a hydraulic circuit for extending and retracting a double-acting hydraulic cylinder with provision for holding pressure, leakage compensation, and power savings. The circuit may include the following essential components: a pump, a check valve, an accumulator, a 4/3-DC valve and a double-acting cylinder.
41. A large hydraulic cylinder is to clamp a workpiece. The cylinder requires a high volume of fluid while it is extending rapidly. However, during the period while the cylinder is clamping, the circuit needs no additional fluid. Assume that the circuit has extended dwell time. Develop a circuit that is capable of storing energy at times of low demand and use the stored energy when a large volume of the fluid is required for a short period.
42. A diesel engine is started using a hydraulic motor. The maximum power requirement from the engine is only for a short time when the start signal is given with a 2/2-DC valve and an interval between two starting operations is long. An accumulator is used to supplement the pump power during the starting operation. The pump should unload during idle time. Develop a hydraulic circuit to implement the scheme.

17 | Numerical Problems

1. Find the size of a gas-charged accumulator used to supply 7370 cm³ of fluid to a hydraulic cylinder with a maximum pressure of 186 bar and a minimum pressure of 126 bar. Assume that the accumulator is used for slow charging and slow discharging. The pre-charge pressure is 120 bar. Assume negligible temperature variations. [Ans: 24.1 litres]

2. A gas-charged accumulator charges and discharges quickly at the rate of 3.3 litres in 3 seconds. It has a minimum to a maximum working pressure range of 90 bar-180 bar and an operating temperature range of 20 °C-80 °C. Assume the pre-charge pressure as 90% of the minimum pressure. Calculate the accumulator capacity. [Ans: 11 litres]

3. A gas-charged accumulator must discharge 5 litres of fluid in 3 seconds with a change of pressure from 280 bar to 208 bar. The loading time is 5 minutes. Define the capacity keeping in mind that the ambient temperature changes from 20°C to 80°C. Assume the adiabatic and isothermal correction factors for higher pressures C_a as 0.75 and C_i as 0.75, respectively.

Objective type questions: Answer key
1-b, 2-b, 3-c, 4-b, 5-d

Appendix 1

Specifications of Piston Accumulators

Table A1.1

Gas capacity		Fluid capacity		Maximum working pressure	
cc	in³	litre	Gallon	bar	psi
226	14	0.22	0.0625	345	5000
246	15	0.24	0.0625	207	3000
251	15	0.25	0.0625	690	10000
257	16	0.24	0.0625	172	2500
493	30	0.49	0.125	690	10000
498	30	0.49	0.125	345	5000
501	31	0.5	0.125	172	2500
524	32	0.5	0.125	207	3000
675	60	1	0.25	690	10000
991	61	1	0.25	172	2500
1000	61	1	0.25	345	5000
1016	62	1	0.25	207	3000
1900	116	2	0.5	345	5000
1901	116	2	0.5	207	3000
1910	117	2	0.5	690	10000
2163	132	2	0.5	172	2500
3048	186	3	0.75	345	5000
3868	236	4	1	345	5000
3966	242	4	1	207	3000
3992	244	4	1	690	10000
4080	249	4	1	172	2500
5779	353	6	1.5	345	5000
5867	358	6	1.5	207	3000
5914	361	6	1.5	690	10000
5965	364	6	1.5	172	2500
7663	468	7	2	345	5000
7751	473	8	2	207	3000
7825	478	8	2	690	10000
7866	480	8	2	172	2500
9548	583	9	2.5	345	5000
9652	589	10	2.5	207	3000
9678	591	9	2.5	690	10000
9750	595	10	2.5	172	2500
11432	698	11	3	345	5000

11537	704	12	3	207	3000
11647	711	11	3	690	10000
15412	941	15	4	690	10000
19271	1176	19	5	207	3000
19292	1177	19	5	690	10000
19385	1183	19	5	345	5000
28743	1754	28	7.5	207	3000
28819	1759	28	7.5	345	5000
28849	1760	29	7.5	690	10000
38115	2326	38	10	690	10000
38198	2331	38	10	207	3000
38253	2334	37	10	345	5000
47687	2910	47	12.5	345	5000
47819	2918	47	12.5	207	3000
57054	3482	57	15	690	10000
57121	3486	56	15	345	5000
57266	3495	57	15	207	3000
66555	4061	66	17.5	345	5000
66713	4071	66	17.5	207	3000
75989	4637	75	20	345	5000
75993	4637	76	20	690	10000
76160	4648	76	20	207	3000
90581	5528	86	23	207	3000
94938	5793	94	25	345	5000
95148	5806	95	25	207	3000
113806	6945	113	30	345	5000
114042	6959	114	30	207	3000
151379	9238	151	40	345	5000
151831	9265	151	40	207	3000
189619	11571	189	50	207	3000
191789	11704	189	50	345	5000
230246	14051	228	60	345	5000
231434	14123	227	60	207	3000
267312	16312	265	70	345	5000
269195	16427	265	70	207	3000
305537	18645	303	80	345	5000
306957	18732	303	80	207	3000
343298	20949	341	90	345	5000
344950	21050	341	90	207	3000
381292	23268	379	100	345	5000
382712	23355	379	100	207	3000
584045	35640	568	150	207	3000
773316	47190	757	200	207	3000

Appendix 2

Specifications of Bladder Accumulators

Table A2.1

Gas capacity		Fluid capacity		Maximum working pressure	
cc	in³	litre	Gallon	bar	psi
1196	73	1	0.25	207	3000
3851	235	4	1	207	3000
9454	577	10	2.5	345	5000
9832	600	10	2.5	207	3000
18858	1151	19	5	345	5000
19714	1203	19	5	207	3000
35095	2142	38	10	345	5000
37018	2259	38	10	207	3000
41541	2535	42	11	207	3000
53413	3260	57	15	345	5000
56372	3440	57	15	207	3000

Appendix 3

Specifications of Diaphragm Accumulators

Table A3.1

Nominal volume		Maximum working pressure	
litre	Gallon	bar	psi
0.075	0.02	250	3625
0.16	0.04	210	3000
0.16	0.04	300	4350
0.32	0.08	100	1450
0.32	0.08	210	3000
0.32	0.08	300	4350
0.5	0.13	160	2320
0.5	0.13	210	3000
0.6	0.16	330	4780
0.6	0.16	350	5000
1	0.26	200	2900
1	0.26	330	4780
2	0.53	210	3000
2	0.53	250	3625
2	0.53	330	4780
2.8	0.74	210	3000
2.8	0.74	330	4780
3.5	0.92	210	3000
3.5	0.92	330	4780
4	1	50	725
4	1	180	2600

Appendix 4

A | Specification Parameters, Accumulators

Table A4.1

Sl No	Parameter	Types/ selection / Examples (Typical)
1	Accumulator type	<ul><li>Piston</li><li>Bladder</li><li>Diaphragm (Weld type / Screw type)</li><li>Bellows</li></ul>
2	Nominal volume	Piston: 0.95, 1.89, 3.79, 5.7, 7.6, 9.46, 18.9, 28.4, 37.9 litres Bladder: 0.7, 1.1, 1.5, 1, 2, 2.5, 3.7, 4.9, 5.7, 9.2, 10, 12, 17.8, 18.4, 24.5, 32, 35, 48.5, 54, 60, 85, 105, 133, 170, 201 litres Diaphragm: 0.075, 0.16, 0.19, 0.35, 0.45, 0.5, 0.65, 0.7, 1.1, 1.4, 2, 2.55, 2.8, 4.1, 10.19 litres
3	Flow rate	38, 95, 150 lpm
4	Effective gas volume	
5	Gas compression ratio	4:1, 6:1, 8:1, 10:1, unlimited
6	Maximum working pressure	100, 140, 160, 210, 250, 330, 350, 400, 450, 500, 750 bar
7	Shell material	Carbon steel, stainless steel, carbon steel with a protective coating, low-temperature steel
8	Diaphragm / Bladder / Seal material	<ul><li>Buna N (Acrylonitrile butadiene rubber) -15°C to +80°C</li><li>Low temp Buna N (ECO) (Ethylene oxide epichlorohydrin rubber) -40°C to +80°C</li><li>Viton (FKM) (Fluorine rubber) -15°C to +100°C</li><li>Butyl rubber (IIR) -15°C to +80°C</li></ul>
9	Corrosion protection	<ul><li>Uncoated</li><li>Coated Epoxy coated, Nickel/Chrome Plated (Inside surface)</li></ul>
10	Accumulator model (For bladder type)	<ul><li>Top repairable</li><li>Bottom repairable</li></ul>

Continued…

11	Fluid port size	
12	Fluid port material	Carbon steel, stainless steel, carbon steel with a protective coating, low-temperature steel
13	Fluid port connection	• SAE O-ring • Thread to ISO 228 (BSP) (G¾ to G2½) • Thread to ISO 965/1 (Metric) • Thread to DIN 13 • Thread to ANSI B1. Thread to ANSI B1.20.1 (NPT) • EN 1092-1 welding neck flange • Flange ASME B16.5 • SAE flange 3000 psi / 6000 psi
14	Gas port type	Sealed type / rechargeable type
15	Gas port size	
16	Gas port connection type	• M28 x1.5, M16x1.5, M14x1.5 • M50x1.5 • 5/8- 18 UNF • Not refillable (screw / welded)
17	Gas charge kit	
18	Operating temp	-45 to 160°C
19	Country of destination	
20	Shell certification/ Compliance standard	ASME, TUV, PED (CE), CRN, AS1210
21	Mounting support	(eg. Clamps and bracket)

B | Essential Specification Parameters, Safety Blocks

- Size (eg. NG 6, NG10, NG20, NG32)
- Maximum operating pressure (eg. 350 bar)
- Shut off valve, poppet type
- Bleed valve, poppet type (Manual, Manual + Solenoid)
- Pressure relief valve Rating (eg. 250, 350 bar)
- Gauge
- Seal (eg. Nitrile)
- Type of connection (eg. Piping connection, sub-plate connection)
- Port sizes (G½ to G 1½)
- Accumulator-side connection type [eg. SAE thread (ANSI B1.1)]
- Pressure port and gauge port connection types
- Tank port connection type
- Fluid compatibility (eg. Mineral-based)
- Temperature rating

Appendix 5

Safety Standards for High-pressure Vessels

The design, manufacture, and use of the accumulator should conform to various national and international standards. Some of the standards of pressure vessels (accumulators) are given below.

Table A5.1

Standard	Applicable region or country
ASME BOILER and Pressure Vessel Code (American Society of Mechanical Engineers)	U S A
Pressure Equipment Directive - PED	Europe, including the UK
Reg. 365 B: Steam receivers, Separators, catch waters, Accumulators, and similar vessels under Indian Boiler Regulations	India
CSA B51, Boiler, pressure vessel, and pressure piping code, Canadian Registration Numbers (CRN)	Canada
Regulatory Rule NR-13, Brazilian Registered Engineers (BRE)	Brazil
AS1210, Pressure vessels, Standards Australia	Australia
High-Pressure Gas Safety Law (high-pressure gas production equipment)	Japan
Industrial Safety and Health Law (class-2 pressure vessel)	Japan
Regulation for Boiler and Pressure Vessel Manufacture Licensing	China
GOST standards	Russia
American Bureau of Shipping (ABS)	Shipping vessels and oil rigs
Off-shore Standard DNV-OS-E101, Det Norske Veritas (DNV)	Oil and gas applications

18 | References

1. Article on 'Advice For Maintaining Hydraulic Accumulators' by Brendan Casey, Machinery Lubrication (7/2009), by Noria
2. Article on: 'Hydraulic accumulators, Book 2, Chapter 1, Part 1, Part 2, Part 3', Hydraulics & Pneumatic Magazine.
3. Article on: 'What is an accumulator?' & TOBUL_Int_ Catalog_101912v2, Tobul Accumulator, Inc., USA.
4. Catalogue on: 'Accumulators V-FIFI-MC003-E' July 2005, EATON, Eden Prairie, MN, USA, www.hydraulics.eaton.com
5. Catalogue on: 'Sizing and Selection – Piston Accumulators, Bladder Accumulators, Kleen Vent', Catalog No. HY10-1630/US, Parker Hannifin Corporation, Hydraulic Accumulator Division, Illinois USA.
6. Catalogues on 'Hydro-pneumatic accumulators', 'Precautions for use and maintenance recommendations', 'energy, silence, comfort, service life...', 'What are accumulators used for?', HYDRO LEDUC, FRANCE, www.hydroleduc.com
7. Document on 'Accumulators and Reservoirs', Senior Aerospace Metal Bellows, USA
8. Document on 'Accumulators', Bosch Rexroth Corporation, Industrial Hydraulics Division, USA
9. Document on 'Accumulators', Catalog V-FIFI-MC003-E July 2005, EATON Hydraulics, USA
10. Document on 'Bladder-type accumulator' RE 50170/02.12, Bosch Rexroth AG, Hydraulics, Germany
11. Document on 'Heavy Diesel Engines Metal Bellows Accumulators' HYDAC International, Germany
12. Document on: 'Accumulator sizing' EPE ITALIANA Srl, COLOGNO MONZESE (MI).
13. Document on: 'Accumulator station, RE 50135/07.11', Bosch Rexroth AG, Hydraulics, Germany
14. Publications Department of Womac Machine Supply Company, Fluid Power In Plant and Field, Second Edition
15. Vocational Training Course, HYDRAULICS – 21 Exercises with Instructions, published by Bundesinstitut fur Berufsbildungsforschung, Berlin, 1973

Fluid Power Educational Series Books

1. Pneumatic Systems and Circuits -Basic Level (In the SI Units)
2. Industrial Pneumatics -Basic Level (In the English Units)
3. Pneumatic Systems and Circuits -Advanced Level
4. Electro-Pneumatics and Automation
5. Design of Pneumatic Systems (In the SI Units)
6. Design Concepts in Pneumatic Systems (In the English Units)
7. Maintenance, Troubleshooting, and Safety in Pneumatic Systems
8. Industrial Hydraulic Systems and Circuits -Basic Level (In the SI Units)
9. Industrial Hydraulics -Basic Level (In the English Units)
10. Hydraulic Fluids
11. Hydraulic Filters: Construction, Installation Locations, and Specifications
12. Hydraulic Power Packs (In the SI Units)
13. Power Packs in Hydraulic Systems (In the English Units)
14. Hydraulic Cylinders (In the SI Units)
15. Hydraulic Linear Actuators (In the English Units)
16. Hydraulic Motors (In the SI Units)
17. Hydraulic Rotary Actuators (In the English Units)
18. Hydraulic Accumulators and Circuits (In the SI Units)
19. Accumulators in Hydraulic Systems (In the English Units)
20. Hydraulic Pipes, Tubes, and Hoses (In the SI Units)
21. Pipes, Tubes, and Hoses in Hydraulic Systems (English Units)
22. Design of Industrial Hydraulic Systems (In the SI Units)
23. Design Concepts in Industrial Hydraulic Systems (English Units)
24. Maintenance, Troubleshooting, and Safety in Hydraulic Systems
25. Hydrostatic Transmissions (HSTs) (In the SI Units)
26. Concepts of Hydrostatic Transmissions (In the English Units)
27. Load Sensing Hydraulic Systems (In the SI Units)
28. Concepts of Load Sensing Hydraulic Systems (English Units)
29. Electro-hydraulic Proportional Valves
30. Electro-hydraulic Servo Valves
31. Cartridge Valves
32. Electro-hydraulic Systems and Relay Circuits
33. Practical Book: Pneumatics - Basic Level
34. Practical Book: Electro-pneumatics - Basic Level
35. Practical Book: Industrial Hydraulics – Basic Level
36. Programmable Logic Controllers and Programming Concepts

For more details, please visit: **https://jojibooks.com**

About the Author

Joji Parambath is a trainer in the field of Pneumatics, Hydraulics, and PLC, for over 25 years. During his career, he trained numerous professionals from the industries as well as faculty members and students of engineering institutions.

At present, he is the key trainer at Fluidsys Training Centre, Bangalore, India, (https://fluidsys.org) which is providing training in the field of Pneumatics and Hydraulics. He has already written two books on Pneumatics and Hydraulics. The publication of the present series of 36 books is intended to restructure and update the existing books.

The author wishes to thank all trainees for their lively interaction and many useful suggestions during the training programmes that prompted the author to write the present series of books. You may send your feedback to joji.p@hotmail.com

10th June 2020